SUCCESS IN ORGANIC CHEMISTRY

Learn Tips. Learn Basics. Make an A!

By Paris Hamilton, Ph.D.

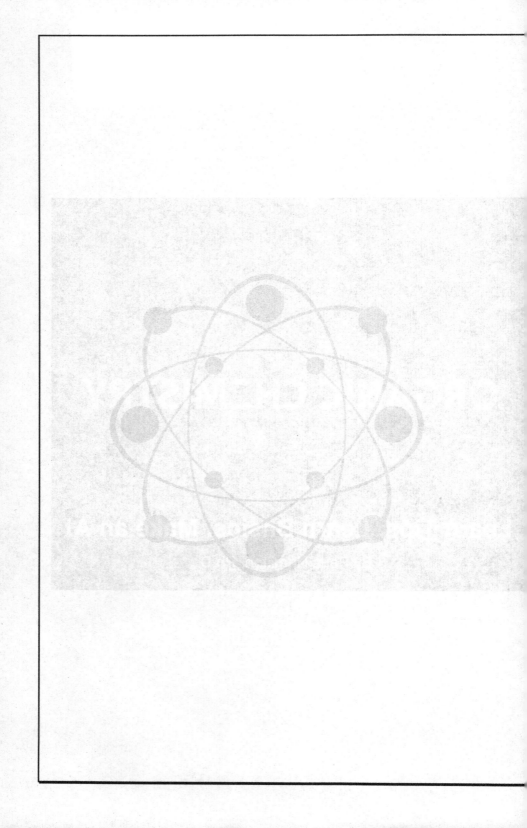

SUCCESS IN ORGANIC CHEMISTRY
By Paris Hamilton, Ph.D.

Copyright © 2016-2017 Paris Hamilton, Ph.D.
All Rights Reserved

Please direct all inquiries to hamiltpl@gmail.com

The information contained in this guide is for informational purposes only.

Disclaimer: You should always seek the advice of a professional before acting on any of the information contained in these materials.

No part of this publication shall be reproduced, transmitted, or sold in whole or in part in any form, without the prior written consent of the author. All trademarks and registered trademarks appearing in this guide are the property of their respective owners.

There are no warranties expressed or implied about any of the procedures described in this work including any warranties of merchant ability and fitness for a particular purpose. Use of any information in this publication is at the risk of the reader, and the publisher and author assume no responsibility for any damages as a use of this material.

If you have questions about this book, feel free to email or tweet me @ParisLHamilton.

Preface

To me, the way that traditional organic chemistry courses are taught today makes them seem like they are preparing students for careers of yesterday. From what I've heard from most students, they just don't see how they will use the knowledge taught in class within their everyday lives.

One particular semester I was tutoring three girls (amongst others). One girl was getting private tutoring, but the other two girls were meeting with me at the same time. All three girls had the same professor for organic chemistry so they had the same assignments. The difference between these girls came in how they approached the material. Two of the girls couldn't care less about actually learning organic chemistry; one of them just wanted to remember enough to pass the class so she could go on to veterinary school. However, the third girl was different. Even though the material was completely new to all three girls upon it being introduced, the third girl seemed to pick up on the material more quickly than the other two. Her ability to see chemistry as it related to her life sparked a curiosity within her that seemed to make her learning process easier. The other two struggled with the material so I had to go over the topics multiple times, in different ways, until some light bulb magically went off in their head and they realized the pattern(s) that I was showing. The first two girls had to put forth way more effort than the third girl to get good grades, but I think their main struggles could have been avoided.

Preface

The intent of this eBook is to shed light on things that hinder a lot of students while they are trying to learn organic chemistry. Additionally, I hope to depict some places that you might encounter chemistry on a regular basis. It is my hope that in reading and understanding these common traits among students, new students taking the course will know some steps to take in becoming a better and more active learner. Hopefully acting on this information will make your learning experience in organic chemistry a tad bit easier.

I would be mistaken if I did not mention other books that seemingly deal with this same topic. A very good book is Organic Chemistry as a Second Language by David R. Klein. That book is geared towards helping you study more efficiently throughout your time in the organic chemistry course. It essentially expands upon what I discuss briefly in part three of this book.

A book like Organic Chemistry as a Second Language is great for learning how to practice working through assigned problems and it gets very technical with the level of chemistry. I absolutely would recommend it to anyone who wants to apply the concepts in this book to detailed examples.

Preface

The goal of my book is not intended to dive into technical jargon, but break concepts down to a point that you can understand so that you have a foundation upon which to build a growing knowledge upon. This book is about developing the right mindset before you embark on the path of studying and as such it is meant to be an easy read.

To all of my readers, I just want to say thank you for the gift of your support. I really do hope this information guides you to a new way of thinking.

Table of Contents

UCCESS IN ORGANIC CHEMISTRY 3

reface 4

.... 5

able of Contents 6

ntroduction 9

ART 1: TRAITS OF SUCCESSFUL STUDENTS 12

1.1– Confidence 12

1.2 – Ability to make connections 13

    1.2.1   ...between what's learned in class to unknown problems 13

    1.2.2   ...between what's learned in class to what is relevant to themselves 16

1.3 – Good memory 18

1.4 – Ability to reflect constructively on what they did 21

1.5 – Ability to visualize in 3D 23

1.6 – Critical Thinking Skills 25

Conclusion 26

ART II: Developing Effective Thinking 29

    Understand Basics 31

    Make Mistakes 34

    Ask Effective Questions 38

    Follow The Flow of Ideas 42

ART III: What You Need To Know About Organic Chemistry 45

3.1 – What is considered an organic molecule? 45

3.2 – How are structures drawn? 46

3.3 – What are "functional groups"? 48

3.4 – Electron attraction 49

3.5 – Some simplified basics 5

3.6 – Making Connections 5

PART IV: EXAMPLES OF ORGANIC MOLECULES IN PLACES THAT YOU
ENCOUNTER DAILY.. 6

EXAMPLE 1: NATURAL BODY STIMULANTS.. 6

EXAMPLE 2: THE PHEROMONE CONNECTION 6

EXAMPLE 3: SUPPLEMENTS 6

EXAMPLE 4: THE CAFFEINE KICK AND OTHER CONSEQUENCES 7

EXAMPLE 5: SUGARS: SINK OR FLOAT? 7

EXAMPLE 6: LETS HIT THE GYM 8

EXAMPLE 7: WEEKEND DRINKING 8

EXAMPLE 8: HAPPY PEOPLE LIVE LONGER 9

A Challenge to You: 9

About the Author 10

Study Guides and Practice Problems: 10

More Quotes That I Like... 10

Introduction

After watching students that have done well in organic chemistry and students that have not done well in the class, one trait (amongst others) stands out. That trait is confidence. Students that enter the class with the mentality that organic chemistry is difficult are ultimately the students who make the class harder than it has to be. They are defeated before they start seeing the material. Confident students more easily see the patterns that emerge throughout the class. They seem better able to apply the concepts they learn in class to problems that they have not seen before.

So, the point of telling you this is so that you take on the organic chemistry course with the mindset that the material being taught is really not as bad as you may have heard. Just because a course was hard for one person doesn't mean that it will be hard (or has to be hard) for you. People put different levels of effort and time into a particular subject so don't adopt someone's way of thinking unless it is beneficial to you doing well in the course. Look for reoccurring patterns and remember that a professor won't typically ask you test questions that they have not shown you how to arrive at the answer for.

Keep a positive mindset that you CAN understand and you WILL understand the material. If you need to say that before going to lecture or before studying, then that is what you should do to keep your confidence level up.

Not all teachers do the best job of getting this across to students so as a student myself (a graduate student) I've decided to try and write a short book that relates organic chemistry to the everyday lives of college students. The truth is that organic molecules and the functional groups that decorate them are all around us. It is one of my goals with this book to point out some of the organic molecules that you might see on a regular basis, but just might not have known about. It is my hope that sharing this information allows you to go into an organic chemistry course and actually have some interest in learning about organic molecules and some reactions that are possible with them.

A lot of students will have to take organic chemistry whether it's for their major or whether it's a prerequisite for a pre-professional school, but give yourself the best edge possible by knowing a little something before the professor tries to confuse you (not always intentionally) with the seemingly boring details. It's easier to retain information when you can make connections with what you already know so let the chapters of this book serve as a bridge between what you know and what you hope to learn.

Take a journey with me and see things the way that I do at times. I am going to try and limit the technical language to a minimum where possible in an attempt to make this eBook easier to read. So, in the final section of this eBook you will read a combination of short essays that I've written based on things I've read about a topic, classes I have taken, things I've seen, and experienced.

Good luck in your future endeavors and I hope you enjoy this eBook.

> *"All wish to possess knowledge, but few, comparatively speaking, are willing to pay the price." ~Juvenal*

PART 1: TRAITS OF SUCCESSFUL STUDENTS

1.1– Confidence

From the standpoint of someone who has taught numerous students, it's obvious that students (on average) just don't get things when first introduced which isn't a surprise because learning organic chemistry is definitely like learning a new language. Every student doesn't fit into the same categories of why they find organic chemistry difficult. A major confidence killer of a lot of students before they even step foot in the organic chemistry classroom comes from past organic chemistry students complaining about how difficult the tests were, how much time they spent studying only to receive a mediocre grade, how boring the lectures are, or something similar. Hearing these stories implants a negative connotation along with organic chemistry and in a sense can doom you before you even begin. My advice for you is simply to believe in ourself; believe you can understand organic chemistry. If you

on't believe initially, then maybe say it to yourself enough over

me so that you eventually begin to believe it.

.2 – Ability to make connections...

> "Learning is more effective when it is an active rather than a passive process." ~Kurt Lewin

.2.1 ...between what's learned in class to unknown roblems

When sitting in class, professors tend to begin each new

opic by introducing everything on very simple cases just so that

tudents can follow the concepts. Typically, after an

ntroduction is shown on very simple cases, a professor may (or

nay not) attempt working through a slightly more sophisticated

roblem. It may look more complex (and it very may well be)

ven though there may have been only a slight modification to

vhat you just went over. It's important that, as a student trying

o understand, you pay attention to how the professor works

hrough these problems in class. There will be no better way to

earn material than to do practice problems!

We, as humans, seem to retain information a lot better when we make mistakes AND subsequently try to figure out why those mistakes were made. You can't just make a mistake and expect to do better the next time around if you don't take time to analyze what happened. If you get lost in class, it will be imperative that you ask for the professor to go back over that concept on the spot. If you are nervous about speaking up in a large lecture class, sitting very close to the front can help a lot. Sitting so close to where the professor is teaching makes it so that you don't have to raise your voice and it's easier on your nerves (if you get nervous speaking in front of people) if the other students are behind you instead of in front of you. Sitting in the front might even make you feel as though you are in a much smaller class even if there are 200+ students registered.

When students try practice problems on their own, they are forcing themselves to make connections between a concept learned in the classroom and an unknown situation. The more practice you have, the easier it is for your brain to make connections because you have exercised those neural pathways so much. The faster that you are able to recognize connections, the quicker you can work through unknown problems. So when it seems that another student might be faster than you at recognizing the answer to a problem, it might be because they have practiced similar problems more than you have. This feature will be valuable during test time because if you finish before time is up, you will have time to double check your answers and fix any things that you may have overlooked.

Two important notes that I want to mention briefly:

✓ You should make an attempt to read a chapter before the professor teaches the lecture(s) on that topic. It will benefit you greatly to already have some familiarity with the material before hearing the professor talk about the topic. This will allow you to spend most of your time in class actively listening instead of writing every word the professor says while partially listening.

✓ Do your best not to fall behind on topics throughout the semester. You will need to work at each topic when they are introduced rather than allow multiple chapters to pass before trying to do practice problems that cover multiple topics. The topics in organic chemistry build onto each other so falling behind one week could make you feel lost the following week so try your best to stay on top of things.

This may sound like a lot, but it's really not that bad (you're probably thinking, yeah right!). On average, a professor will spend more than one lecture period going through a chapter. This means that you can read a chapter at the beginning of a week, go to the class while actively listening and interacting with the professor, and practice problems outside of class for maybe an hour or two. It's important to realize that you don't have to work through practice problems on your own. You can use the office hours of your professor, have a tutor help you, or even work through them with a classmate or multiple classmates. Don't be discouraged if you don't pick up a topic as quick as another person; everyone learns at their own pace!

1.2.2 ...between what's learned in class to what is relevant to themselves

There are also students who sit in class and don't see how what they are learning could possibly ever be used in their life outside of the classroom or in their future profession. A lot of these students just tune out.

One of the ways that good teachers try to get students interested in organic chemistry is to try and make the topics seem relatable to the lives of students. I know of a lecturer that always gives some mention to street pharmaceuticals in his attempts. I know another professor that has his students read chapters from the book "Napoleon's Buttons: How 17 Molecules Changed History" as he goes through different chemistry topics. Why is that not always helpful though? There is a huge difference between "real-world" and "relevant". There are tons of real-world examples that textbooks or professors can give, but that does not necessarily make it relevant to each student. We often make a faulty assumption when we teach, that real-world and relevant are the same thing. What is relevant to one person may not be relevant to another.

I had a conversation with a professor on this very topic and he goes on to say this: "when I was out in Oregon, a couple of colleagues of mine would put together a special topics course called the chemistry of skiing; an incredibly interesting course. It really gets the kids excited about doing chemistry. And so, they came to me and asked how can we put this together and kind of modularly and disseminate this so that other people could use it. And I said okay, what is a kid sitting in like south Florida going to do with the chemistry of skiing? For that matter, what is the kid in Indiana going to do where there is a ton of snow, but no hills? The chemistry of skiing in Oregon works because people ski there and it turns out that everybody in that class has season

17

passes to multiple places. In Oregon skiing is such a big thing." I completely agree with what he was saying with that story.

1.3 – Good memory

A lot of pre-professional schools seem to require students take organic chemistry before entering their school. It seems that they use it as a way of seeing how well a student is at memorizing a lot of information and using that information to solve unfamiliar problems.

Having the ability to memorize pieces of information will definitely be useful because you will need those pieces of information to solve the problems given to you. If you don't think that you have a good memory, then you may have to put in more hours of studying than someone else, but try not to compare your ability to learn to someone else; everyone is different.

"In this age, which believes that there is a short cut to everything, the greatest lesson to be learned is that the most difficult way is, in the long run, the easiest." ~Henry Miller

You can find anywhere online or in a smart phone app where people are telling you the facts in organic chemistry. Whether it's how to draw structures, move arrows, which chemicals cause certain functional group transformations, etc. Everybody tells you what the facts are, but not many people can relate it specifically to your life because nobody knows you like you know yourself. Part of it is that those facts that people teach you have to connect with something that you know for the information to stick with you. Because what we do is seemingly come from out of thin air and throw information at people. And it's known that the best way to remember information is to build upon things that you already know. If you have nothing to tether the information thrown at you to, then it's going to seem difficult to retain the information.

Yes, people are known to have some average short attention span when it comes to sitting in lectures and retaining information thrown at them. And the attention span varies from person to person. Sadly, most professors don't care about that because if you are going to really do something meaningful with your life then you need to be able to hold an attention span longer than what a pigeon can do. Out of most large organic classes, majority of the students are looking to enter into a health profession of some sort. If you are looking to go to

19

medical school, at some point you will do a trauma rotation. At that point, you have someone coming in and in 45 seconds or less they will tell you the entire history of this patient. It's not just that you have to take that information in, you've got to process it so quickly because whatever you decide to do in the following minutes will either put that person in the intensive care unit or in the morgue. So, if you cannot follow a professor talking to you in lecture at some slow pace, what are you going to do in the emergency room?

A clear reality is that if you are a student looking to go into a health profession, you should really notice that you can find organic chemistry a lot in biology. Whether you look at the structure of biological molecules or think about the interactions that biological molecules have with one another, you can find organic chemistry principles.

Students have told me that when they are taking a class for a grade such as organic chemistry, they focus mostly on memorizing facts to regurgitate on exams for that good grade rather than learn for retention of that information. I've been that way before too in certain classes; mainly outside of my major. There are plenty of classes that I have taken and done very well in, but could not tell you a thing I learned in that class. It's sad, but true and it happens a lot more than we tend to think about. The purpose of me even mentioning this is in hopes that you will come to the conclusion that you don't want to be that type of person or learner in general. It doesn't benefit you in the long run and that is not the type of habit that you want to form. If you do that for majority of your classes, then you will get to the end of your college career and your degree will be worth nothing because you haven't retained much new knowledge.

Not making the connections between what you are learning to what you already know will actually make it harder for you. And when you are taking an information-dense course such as organic chemistry, making it even harder is the last thing you should want to do. Taking action on becoming an in-depth learner will probably (and likely) be a hard task, but well worth the end result.

1.4 – Ability to reflect constructively on what they did

You know when people say that you learn by doing? Sometimes that's not always the case and you don't learn by doing. You learn by reflecting on what you did because while you're doing, you don't have time to think. Which is why it seems that people learn more from their mistakes as opposed to things they do correctly because when you make a mistake you take time to dwell on why you did something wrong.

As far as teaching and having students learn stuff it seems that it's more beneficial to have students do some type of group work. However, working in groups would only be effective if done in a useful manner. It should give students a way to talk out what they might be thinking using terminology from the course (not their own shorthand) so that others can understand them and they make sense of their thoughts to themselves as well. On the other hand, if group work is done such that work is divided amongst the members and they work on problems individually and then put all the work together and people double check answers on their own; the group setting won't be beneficial. There something about negotiating through problems with other people and the use of the language that helps cement concepts in our heads as students.

If there is no group work given in your class, which isn't uncommon for large organic chemistry lectures, maybe you can talk through notes with someone else in your class or work on practice problems in your book with someone. Now, like a lot of people, when taking a class I prefer not to study with a group because everybody goes at their own pace and I know that I can breeze through notes faster than some people. However, a helpful way that I have found to study even on my own has been to review my notes one time, close my book, and talk out (to myself and maybe in front of a dry erase board) all of the things I am expected to know about a topic. It never fails that there will be things that I know well and some things that I don't even know where to start explaining. So the things that I know well are things that I don't study as much because they might be a bit more straight-forward than other topics. That leads me to put in a solid two or three hours of studying the things that I couldn't explain at all. Once I finish, I go back to closing my book, and explaining all of the topics to myself even the ones that I claimed to know again. If there are still holes in my knowledge without looking at my notes, then I study more. I find that I really do not understand a topic if I cannot explain it to myself or others without the use of notes. This technique may or may not work for you; everybody is different, but it's proven to work for me as a student.

1.5 – Ability to visualize in 3D

There are those students who have a problem visualizing molecules in three-dimensions. This starts to become a problem around the time that the topic of stereochemistry is introduced. This problem doesn't generally go away once moved on to a new chapter because a lot of reactions that are learned after that topic create stereocenters. My words of advice here are to:

1) look into modeling kits that you can physically build molecules with and rotate around so that you can physically see the arrangement of atoms in space.

2) practice drawing structures that have stereochemistry. This will be important because professors won't always let students bring model kits to an exam. Even if a professor did allow model kits, it would probably take up some of your valuable test time if you had to build and take apart a molecule for each question so the more practice you can get without the use of model kits, the better.

3) look into some video tutorials online that help explaining this topic. At least you can watch someone else work through relevant problems repeatedly until it begins to make sense to you.

) study these problems with a classmate or tutor. Having the bility to talk through problems is really a powerful way of earning because it seemingly forces you to think more in-depth bout topics compared to you just re-reading your notes.

.6 – Critical Thinking Skills

This characteristic is probably one of the traits that I find ɔ be most useful and I think that it goes somewhat with having ome scientific intuition as well.

Yes, organic chemistry is like learning a new language, ut even in a subject like French or Spanish, you will have to nemorize vocabulary before you can string together sentences r speak back and forth with someone in that language. How do ou expect to solve a puzzle, which is what reactions look like, you don't know the pieces that are at your disposal? Not nowing basic pieces is what leads people to not knowing how ɔ start most problems.

As far as scientific intuition, some answers are more easonable than others and it is up to you to determine the best ne. This is especially true in test situations so it's better to ractice your thinking regularly BEFORE you even think you will need to use that skill. It's not something that will be acquired overnight so waiting until the day before exams to study is definitely not the most effective way to pass exams.

In situations when you are given a problem and your first thoughts are along the lines of "where do I even begin?", it's those skills that can at least get you started thinking about what information you need to know to be able to solve a problem. Once you realize the information that you need, you can assess whether you actually know that information or whether you need to study it more.

Conclusion

It seems that people trying to get chemistry across to students have to use a variety of techniques to get students to understand the material. Of course, as students you will have to do some practice problems, studying, and reflecting on your own. And part of it is that ultimately you, as the learner, have to see the utility for what you are learning because if you don't then what the professor or tutor does may not matter.

The way, or the order of topics, that the traditional organic chemistry course is taught has not changed in decades

If you have a parent who went to college and took organic chemistry, then they probably learned the topics in the same way that you are being fed the information. If you look in an organic chemistry lecture, the chances are that majority of the students are some biological science type of major. Yet most professors are not trained to teach in a way that relates the information of the class to who the real audience is. That is part of the failure of the way organic chemistry is mainly taught, that it's not taught in such a way that people who don't want to be chemists can still use the information. Now that you know this, the burden falls on you as the learner to do what is in your power to make the information more digestible.

Some take home points:

- ✓ Practice problems should become your best-friend. The more comfortable you become with seeing different variations to problems of the same topic, the easier it will be for you to answer unknown questions within that topic.
- ✓ The professor or tutor has no way of knowing what is relevant to you specifically so it's up to you to make a connection.

✓ Go to office hours! Having your professor review topics one on one with you can greatly help your understanding and confidence. If you think you know everything, go get confirmation on what you think you know. If you have questions, get them answered by the person who is going to be asking you the exam questions. If you just need clarification on a topic, ask a broad question like "can you go back over S_N2 versus E2 reactions?"

PART II: DEVELOPING EFFECTIVE THINKING

You can read tips on how successful students have succeeded in organic chemistry courses before, but that won't do you much good unless you know how to get to the level that they seemingly mastered. In this section, I want to give you a framework for how to get that confidence that those successful students displayed, how to make learning look a bit easier to other students around you, and how to maximize your efforts and get the best results.

Implementing the following system of thinking will get you to a point where it is no longer about how your professor teaches the subject to you, because you will be making the material engaging for yourself every time you step into the lecture hall. You won't sit and start to yawn uncontrollably through lectures or spend most of the time texting a friend, you will look at the material with a more critical eye and as a result you will retain more information.

I can't count how many times I have sat in a lecture, taking notes, really feeling like I understood the material as the professor discussed it in class only to later re-read my notes and feel a bit lost. Of course, people prescribe different things to get over this. One is to read the chapter (and even take notes) in advance before attending the lecture. That way your focus is more tuned onto what the professor is saying and trying to deeply understand material instead of focusing more on writing every word or example so that you can hopefully teach yourself the material again later. That tactic of reading before class does work, but for students usually taking 15-18 credit hours per semester and most taking other hard classes in order to get into medical (or other pre-professional) school, it's not always the most practical; especially if they want a life outside of class. So, what I suggest is the following, which I will explain a little more after this brief synopsis. 1) Take time to identify and truly understand the basic concepts of the subject. This will save you a ton of time in studying later on even though it will likely take a good amount of time up front to truly master the basics. 2) Make mistakes and learn from failures. This requires that you actually take action, but recognize that it's not just about completing a homework assignment. 3) Ask effective questions. When you are either in class or on your own, be able to ask the questions that really lead you to what you want to know instead of vague ones that leave you clueless. 4) Follow the flow of ideas. Understand that new concepts in the course will arise from previously taught concepts. If you can follow where new ideas came from, then you should also be able to predict, to some degree (even if a small degree), what concepts come next.

Understand Basics

The brain is the body's natural chemist. It releases all kinds of chemicals into your system to deal with various situations. If you're happy, it will release some dopamine. If you're stressed, it will release cortisol. If you are scared, it will release adrenaline. The most important chemist in your world, is your own wonderful brain. Man tries to replicate the laws of nature and even improve upon it by synthesizing variations of nature's molecules in a laboratory setting. The organic chemistry course that you are embarking on represents just one major branch of chemistry that scientists have developed known rules for in order to create new molecules that hope to find some use in society. If you or anyone you know takes medicine prescribed by a doctor or without a prescription off of a store shelf, then you are seeing the end result of organic chemists at work within the pharmaceutical industry. This industry is just one of many that organic chemists play a major role within our society.

You will look at your professors in awe wondering how they could possibly know all of the examples in the textbooks that deal with chemistry. And it's not that they have memorized or know all of the examples. It's that they understand the basic principles that make up all of the problems so they have a better vantage point from which they are viewing things. In any setting or subject, the master always looks at a topic differently than the beginner. The question you should be asking yourself and striving to answer is "what are the most important basic concepts that this subject is built upon?" And I go into those in the next part of this guide so be sure to pay particular attention to that section as well. But for now, realize that understanding the basics of organic chemistry (or any subject) are where you want to begin building your foundation of soon to be deep knowledge.

There will be many times in organic chemistry when you are presented with a problem that you have no idea how to solve. In those cases, try to talk yourself through it. Maybe ask a question like "what did we already learn that looks familiar to this reaction?"

Take time to carefully dissect problems. Break them down into smaller, more manageable problems. Be able to clearly identify the electrophile and nucleophile. Take notice of times when acids and bases are present. Let's say you are given a relatively simple substitution problem with an alkyl halide and a nucleophile, but you are unsure of what the product's stereochemistry will be (i.e. not sure if it reacts via S_N1 or S_N2). First, look at the problem and try to focus on one aspect of it that may help you determine the answer. You could focus on one of many things such as (but not limited to) what is the nucleophile or the structure of the electrophile. If you choose to focus on the structure of the electrophile, okay, that is a good start that you have at least chosen SOMETHING. You can then go on to narrow down how it might react by recalling the reactivity tendencies of primary, secondary, and tertiary alkyl halides. If you chose to begin solving the problem by focusing on the nucleophile, you could direct your attention to whether or not it would be considered a good or bad nucleophile. Whichever way you choose to start answering a difficult problem, always try to break it down to something more simple that you CAN answer. This will often take you back to basics so that you can build from there.

Make Mistakes

The culture of a large part of society, as it relates to dealing with failure, can be viewed in the attitude of students. Most professors punish students for mistakes on assignments and tests by not only telling them something is wrong, but not allowing them to correct their mistakes for redemption of their grade. If someone fails the first time, but later understands their mistake, haven't they grown as a student? Why should their first attempt be permanently stamped onto their identity? It's assumed that because professors tell you that you can do some homework problems out of the book or online, that you will fail on your own outside of class and learn from those mistakes. Not realizing that most students do homework either in their room, on their bed, around friends in the library, periodically checking text messages, listening to music, and maybe even watching television. You may use a textbook, google, a friend, or even a tutor to help you answer your homework problems; it's like you have all of these life-lines to call upon in your time of need. But then, you are given a time-pressured exam that forces you to reenact whatever problem solving method you used in private when you felt like you had all the time and resources in the world. If you begin with the assumption that you may not be trained adequately for exams, then you direct your brain to start thinking of how YOU can better prepare YOURSELF; independent of what the professor may or may not teach you.

Let's think for a moment. When scientists are faced with a challenging problem, do you think they tackle it head on? No. They break the problem down into smaller questions and smaller more manageable problems, that they can solve. When president Roosevelt challenged American scientists to land a man on the moon in the 1960's, their first attempt wasn't to send a man into space. They merely wanted to hit the moon first. They had to solve the problem of how to get onto the moon first before tackling the larger issue of getting a human on the moon.

Here's another illustration in failure. A 25-year-old girl's mother died in 1990, which began a very difficult period in her life. She ended up leaving her job, having a kid, getting a divorce (that later involved a restraining order), and living on welfare. Eventually she became clinically depressed and even contemplated suicide. Two things that kept her going were her young daughter and an idea. Her idea was about a boy who attended a wizardry school. She found the time to write whenever she could and eventually finished the first Harry

35

Potter novel 5 years later. After she found an agent, J.K. Rowling's manuscript was rejected by 12 different publishing houses before someone agreed to publish it. We know how the story ends of course, with a 7 book series, soon to be 8th book being released in July 2016. She landed movie deals and even had theme parks build sections around her characters and fantasy world. She is the first author to become a billionaire from writing. When talking about her success, she mentions owing it to her failures in statements like this: "Failure meant a stripping away of the inessential. I stopped pretending to myself that I was anything other than what I was, and began to direct all my energy into finishing the only work that mattered to me. Had I really succeeded at anything else, I might never have found the determination to succeed in the one arena I believed I truly belonged. I was set free, because my greatest fear had been realized, and I was still alive, and I still had a daughter whom I adored, and I had an old typewriter and a big idea. And so rock bottom became the solid foundation on which I rebuilt my life." J.K. Rowling could have given up at any point on her journey because it was paved with one failure after another, but she didn't and she found a strength within that allowed her to not only learn from her failures, but thrive in the face of them.

The traditional way of teaching anything typically goes as follows: Teacher explains a concept, shows you how to solve it, and then lets you attempt to solve it on your own. More recently, studies in cognitive science have shown that presenting a student with the problem first and allowing them to attempt solving it before offering assistance leads them to understand concepts more deeply and in turn better apply those concepts to new problems. In most cases, insights or those "aha" moments don't arise until AFTER someone has been faced with a psychological "moment of defeat." That is to say, you will likely feel like you have exhausted all of the tools in your mental toolbox on a problem, before reaching a moment when everything becomes clear. It's an essential step not only in solving new problems, but also in developing creativity. So, the lesson here is simply, don't give up!

"Many of life's failures are people who did not realize how close they were to success when they gave up."
-Thomas Edison

When presented with a difficult problem in chemistry, you may not know the correct answer, but you can definitely get it wrong. When you get a problem wrong, don't view that as the end, but as a starting point. Ask yourself "what is wrong with this answer?" Once you've identified one thing that is wrong with the answer, ask "how can I change this one thing to make it right?" Once you've corrected that one thing, ask yourself again "now is the answer right?" If not, identify a new problem and solve that. Keep repeating the process until you have the right answer.

Ask Effective Questions

Force yourself to ask questions on a regular basis and not just when you don't understand something. These questions do not necessarily have to be spoken aloud; they ca be probing questions that you ask yourself internally in an effo to understand why a professor just went over a topic of how th new information relates back to the basics. Whether or not you convert something you hear into something you learn will be determined by your level of active versus passive listening. Listening alone is merely NOT enough when sitting in a class; is the questions or internal dialogue that will keep you interested and engaged in what is being talked about. We have all been in a situation where someone is talking and our minds wander off to think about something else, but that is NOT the headspace you want to be in if you are sitting in a lecture. You just need to get yourself into a mindset of focus and questioning that will last for a 40-90-minute span of time. After which, you can walk out of the lecture hall and into the calming winds of the outdoors with your cell phone in hand. People who question things are far more productive, are more engaged, and consequently are smarter than those who do not.

Questions are equally, if not more, important than the answers they inspire. Why is that? Because the question helps you focus on specific aspects of a problem. Unfortunately, most people spend their lives asking and focusing on the wrong questions. So that begs the question, what are the right questions? The right questions are those that highlight some feature of a problem or hidden assumptions and lead you into what to do next. Let's look at some examples.

Let's first go through what type of questions are NOT considered effective questions.

1. How can I get an A in organic chemistry? That's a relatively easy question to answer. The answer is either, 1) end up with an average over 90% or 2) get an A on majority of the exams and assignments.

2. How can I memorize more? Are you making a hidden assumption here? Maybe you are assuming that memorizing is the best way to succeed in organic chemistry. (Hint: it's not) Being able to recall information, or memorizing, is only the basic level of knowledge. It is how you use that knowledge to solve new problems that is more important. Majority of student's taking organic chemistry will never memorize all examples and practice problems so asking "how can I memorize more?" is not an effective question.

And now let's look at some more effective questions.

1. How can I become a more effective thinker in organic chemistry?
2. How can I be more engaged in the course material?
3. How can I increase my curiosity about chemistry?
4. Could I give a lecture explaining the material or teach it to someone?
5. What basic principles can I tie all problems back to so that I don't have to memorize everything?

As you see from these types of questions, they lead you to specific actions that you can take. Whether its joining a study group, getting a tutor, or even better, finding a study partner to teach concepts to. So, when you are in lectures or doing homework, even more so when you find yourself stuck on something, ask yourself constantly if you are asking the right question. If not, recognize it and frame new ones. Your goal should NOT be to memorize the book, BUT INSTEAD to develop an effective way of thinking about the material so that you can apply memorized facts to unknown and new problems.

Follow The Flow of Ideas

This tactic relates to the lesson of understanding the basics. It's like the cliché saying of "to know where you're going, understand where you've been." Every time you learn a new subject or reaction in class, relate it back to the basic ideas that you understand. Really examine how the problem that you are presented with stems from those starting points and bride any gaps that you find. When you can logically follow where ideas flow from, you will be more equipped to understand where those ideas may take you in the future. Your professor understands how each idea flows into the next which makes the subject seem much easier to them. They might not necessarily teach you how to make the connections during the lecture setting so it is up to you to make the necessary connections. Now, even though they don't go through this in the lecture, that does not mean that they won't go through it with you alone if you went to office hours and asked them to explain how one idea flows into the next. I've found that students often feel like their professor is much more helpful during office hours when you can get private explanations. That is likely due to multiple reasons such as the professor trying to teach hundreds of personalities at one time or even the professor trying to stay on

heir semester schedule with what lecture to cover for that
eek.

How might this help you? Not only will you get a better
ense of the overall picture of organic chemistry, but when
reparing for exams, you should be able to write your own
ractice questions. The practice questions you write should be
ased on the main concepts that you learned and allow you to
ractice different variations of those concepts. That's
ssentially what an exam is testing you on; variations of
ndamental concepts. Most students get tripped up or
onfused when they think they know something, but the
rofessor throws in one slight variation. That variation does not
nange the fundamental concepts, but it looks unfamiliar to the
ntrained eye and thus makes a student feel lost.

No matter where you are (or think you are) in your
nderstanding, your understanding can still be improved. Don't
ettle for merely trying to memorize an isolated fact. Take any
entence over twelve words long (twelve is an arbitrary length)
nd jumble up the words into a random order. Wouldn't it be
ore difficult to memorize each word on the list if it were out of
s logical order? Don't put the cart before the horse. Strive to
arn the foundations of ideas and then build upon those. With
l things considered, it is not hard for you to learn these four

43

lessons to develop effective thinking, it just requires that you g[o] out and do it. You can learn anything in the moment, but unles[s] you practice what you learned, it won't stick for long. If you car[n] implement these tactics in your learning of organic chemistry, I know that you will be successful.

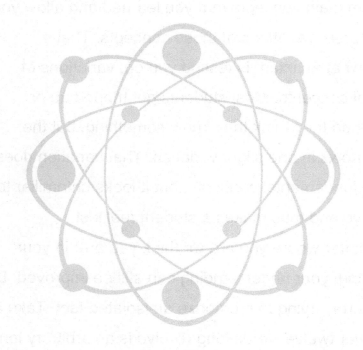

PART III: WHAT YOU NEED TO KNOW ABOUT ORGO

This section isn't all inclusive in the sense that it's not ALL you need to know about organic chemistry to pass the course; more like some minor specifics to comprehend that will help you understand more complicated information.

3.1 – What is considered an organic molecule?

Organic chemistry deals with the chemistry of carbon containing molecules. These molecules are all around us and ever present in nature. In fact, a great number of organic chemists gain their inspiration from nature itself. Not only could this information be useful to students who will take organic chemistry (or already are in the course), but to individuals that might want to see their world a little differently.

is the same as

Random fact – Electrons are credited as being discovered in the late 1800's; between 1896 to 1898 to be more precise. That is around the first publication time of the famous Dracula story (1897).

3.2 – How are structures drawn?

In chemistry, we use a range of symbols and notations in order to describe how processes are occurring and to display structures.

A small note about understanding the organic molecule structures that you will see within this book and in chemistry courses; not all of the carbon or hydrogen atom symbols (C or H, respectively) will be drawn out explicitly in a molecule. Atoms are connected to one another by sharing two electrons; drawn as a line between two atom symbols (usually one or two letters depending on the element). If you see a line drawn between two points in a molecule, then the lines represent a bond between two atoms.

I would make an attempt to relate the bond between two atoms to a bond between two individuals such as in a marriage, but some atoms form more than one bond and I don't want to relate these structures to polygamy. I guess the bonds between atoms could be more like dating because the bonds can be

46

broken under the right conditions, but I will leave these comparisons alone.

The vertex between two line segments represents a carbon (unless otherwise noted with a letter (or letters) other than 'C').

Carbon is stable with four bonds so if a carbon is only bonded to one other visible carbon (or oxygen, or nitrogen) with a single line, then there are three hydrogen atoms bonded to that carbon. If carbon is bonded to two separate carbons with a single line to each, then that central carbon has two bonds to hydrogen (even though they won't be drawn in) and so on.

You will see bonds between two atoms sometimes drawn as either a solid black wedge or a dashed wedge. Don't let that change throw you off because it is just a way of showing the arrangement of atoms in three-dimensional space. A solid black wedge is telling you that whatever is attached is pointing towards you (out of the plane of the paper or screen where it is drawn). A dashed wedge is telling you that whatever is attached to that bond is pointing away from you (or into the plane of the paper or screen where it is drawn).

3.3 – What are "functional groups"?

In organic chemistry, most of what you learn revolves around recognizing common arrangements of atoms known as "functional groups". These functional groups have their own reactivity and lead to specific products when introduced to particular chemicals. A typical organic chemistry course deals with identifying common functional groups and the different reactions that can happen at those sites. On a daily basis I am sure that you encounter these same functional groups in your everyday life and just may not have known it. Some examples of common functional groups that are very common are hydroxyls (alcohols, -OH), alkenes (or olefins, -C=C), carboxylic acids, (-COOH), and amines (ex. -NH$_2$).

Along with each functional group, you will be told a specific set of chemicals that perform specific reactions. A part of your problem solving skills will come from being given an unfamiliar molecule and recognizing how you can make it in steps using the reactions that you will have learned. It's not always obvious which reactions you will use or even where to begin. There is no way around the fact that you will have to do some memorizing to get a good grade in the class so potentially making a connection to where you encounter these functional groups in your life could make memorizing seem a tad easier.

.4 – Electron attraction

^{6}C ^{7}N

Carbon   Nitrogen

^{8}O ^{9}F

oxygen   fluorine

Atoms have different affinities for attracting electrons towards themselves; in organic chemistry, you will hear this described as the term electronegativity. Let's consider only the four elements that I have drawn above from the periodic table. From left to right (in this row on the periodic table), each element has more protons in its nucleus meaning more of a positive charge. The increase in positive charge of the nucleus, leads to an increased attraction to nearby electrons (which are negatively charged). So let's think about some simplified cases.

C–C   **C–N**   **C–O**   **C–F**

Case #1   Case #2   Case #3   Case #4

In case #1, since carbon is bonded to carbon and both atoms will have the same affinity for electrons, the electrons will be distributed fairly evenly between the two.

In cases 2-4, carbon has a bond to nitrogen, oxygen, or fluorine. Each of those atoms bonded to carbon has a higher affinity for pulling electrons towards itself and away from carbon. Since this is the case, the electrons in the bond between each of the two atoms in these cases will be distributed unevenly; leading to the existence of a partially negative charge on N, O, or F because that's where the electrons are attracted.

3.5 – Some simplified basics

Organic chemistry courses break down into certain particular topics or ideas that get used throughout the first and second semesters; identification of nucleophiles (nucleus-loving), electrophiles (electron-loving), acids, bases, and the use of mechanisms. A nucleus houses protons which are positively charged so a nucleophile has a full or partial negative charge. Electrons are negatively charged so electrophiles are fully or partially positively charged.

In science, opposites attract; meaning a positive charge will interact with a negative charge. The interaction could be covalent or ionic. That is a simplified understanding and in an organic chemistry class things tend to seem more complicated once new terminology is introduced. The first example that comes to mind with this concept is that of two magnets. The magnets have a north pole and a south pole end. If you put the north pole end of one magnet with the south pole end of another magnet, you will get attraction. The closer the two ends are, the stronger the attraction. Two north pole (or two south pole) ends brought together will repel each other.

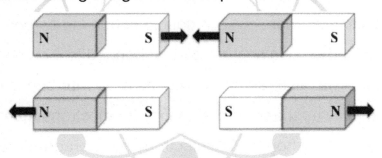

This idea can be applied to molecules found in organic chemistry if you think of the magnet as being divided into a positive (+) side and a negative (-) side; which I'm not saying is correct in reality, but it's easy to visualize. Many organic molecules will be made up of more than just carbon bonded to other carbon atoms; oxygen, nitrogen, bromine, chlorine, and others are very often shown bonded to carbon as well. Majority of the atoms bonded to carbon, attract electrons (which are negatively charged) towards them more so than carbon attracts them; leading to partial (δ) charges arising on atoms in a molecule.

This particular ketone is known as acetone also known as fingernail polish remover

In this case above, I have drawn what is known as a ketone functional group; oxygen double bonded to carbon and that central carbon has another carbon bonded to it on both sides. Notice that all three carbon atoms have four bonds to something. Oxygen has a higher affinity for electrons than carbon, causing carbon to lose electrons on average. When electrons are taken away from a neutral atom, it causes that atom to have a partially positive (δ+) charge or a full positive (+) charge; the former in this case is drawn above.

Now that you understand that concept, if another molecule is mixed with this ketone structure, and that second molecule has a negative charge then the two will react. The position of partial positive charge will form a bond with the atom bearing the negative charge on the other molecule (in simple cases where the negative charge is on a molecule that acts as an attacker and not as a hydrogen taker). And since carbon is only stable with having four bonds, one of the two bonds to oxygen will break because you will learn that a double bond is not energetically favorable as a single bond.

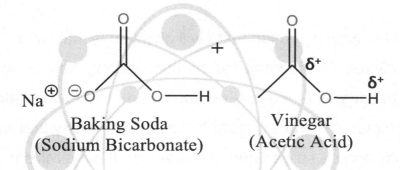

Another example of this negative charge being attracted to a positive charge is in the form of an acid reacting with a base. In theory, the oxygen atom with the negative charge drawn on baking soda (sodium bicarbonate) has two atoms on vinegar (acetic acid) with which it can react. In reality, the hydrogen atom on vinegar reacts fastest for different reasons.

The first reason that I will give you is that vinegar is considered an acid, a weak acid but still an acid. This means that it is a donor of a hydrogen atom. The double arrow that is drawn below means (in the context that I'm using) that the structures can interconvert between the two structures that are shown.

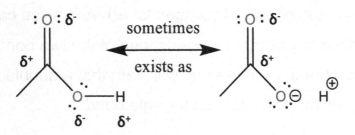

The second reason that I will give you is that when vinegar loses that hydrogen atom, the resulting compound is stable because the negative charge can move between atoms on the molecule. You will learn in organic chemistry that when electrons can be spread over more atoms, the overall structure is more stable than a structure that has electrons (or a negative charge) localized on a single atom.

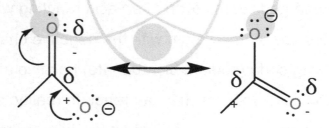

If you notice, there is a full negative charge on oxygen right next to a carbon with a partially positive charge. As I've mentioned before, opposites attract so there is no good reason

54

why that oxygen and carbon atom cannot form a bond together; so, they do. You can see that when that happens the same base is present in the product which means these structures will exist in both forms and just switch back and forth.

The arrows drawn directly on the molecule will be important to recognize later when you are in the class; they describe the movement of two electrons per arrow. The side of the arrow with no head indicates the source of electrons and the side of the arrow with the head describes where those electrons move to. So, if an arrow begins on an atom and goes to the bond between two atoms, a second bond is forming between those two atoms. If an arrow begins on a bond between two atoms and moves to one atom, then one bond is broken between the two atoms and the two electrons that made up that bond will move to the atom that the arrow head points to. Reread those last two sentences if it was confusing at all and compare it to the picture that is drawn above to make sense of what is being said.

Keep some important things in mind when learning this new information. First, you won't always understand a topic when you are first introduced to it. Don't become afraid when this happens, step up to the challenge. Second, the sooner you identify what it is that you don't understand, the sooner you can

55

work on finding answers. Third, the best way to really grasp these topics is to work through as many practice problems as you can. You won't always know the answers to practice problems, but it is better to get them wrong while practicing than to get questions wrong during test time. I heard an entrepreneurial speaker give a lecture one day and when he said "the faster you fail, the faster you get to success" it really stuck with me and you should really keep that in mind as well. Fourth, if you are a full-time student, you will have other classe at the time of taking organic chemistry so traits like organizatio and time management become very important. Procrastinatior will be one of your biggest enemies because falling behind on your understanding of a topic over weeks could lead you to piling on the amount of work you have to do when you finally get around to studying or doing practice problems. Try to set aside at least a couple of hours a week to dedicate to organic chemistry material. If this sounds like a lot, just realize that many people have made an 'A' in this class and so can you. Believe in yourself and put forth your best efforts.

"The faster you fail, the faster you get to success."

3.6 – Making Connections

I was reading a March 2016 article in chemical and engineering news, a popular magazine read by chemists, and it was about how difficult organic chemistry is to students and how it has been for decades. In the article, it discusses how the community of chemists is split on the topic of their being a crisis in teaching organic chemistry. Some people believe that since there are so many online resources available these days such as websites that provide study guides, summary guides, and a bank of practice problems, that there is no real crisis in teaching organic chemistry to students. Others believe that just because there is a wide array of products out there that help supplement the professor, that doesn't equate to no crisis in chemistry education.

Organic chemistry textbooks have been unchanged over the last 50 years so if you are taking the course now, you are essentially learning it the same way and same order that your parents would have learned the material. That is a problem. The way this subject is taught should evolve with the times, especially since the difficulty with teaching it to students has been known for years. More university majors now have to take organic chemistry to fulfill degree requirements so it's no longer just the students who want to go to medical school or another professional heath school. Students from food science, packaging science, and nutrition (amongst others) also have to take the difficult course and may not find the same connections of chemistry to society interesting. It has to be taught so that the widest number of people understand the concepts being taught and that is currently not the case. The issues that students have these days with the subject are the same that students of the past had with this same subject. They don't find it any easier to digest the information nor do they find it any easier to understand the chemical symbols or chemical reactivity. What textbooks try to do these days are to attach some section with biologically relevant molecules; typically, towards the later chapters of the textbook. I think that is one step in the right direction, but I do not think that is a "cure" for issues facing teaching organic chemistry. The relevant examples gives individuals something to connect information to. The issue there is that there are no one or two examples that will resonate with the majority of students taking the course. It may help some, but not all. In this sense, amongst others, organic chemistry textbooks are mostly interchangeable; it doesn't matter a great deal who wrote the book because the information will likely be 90% similar to another book on the same subject.

Now, one of the things that I think would help is if there was a way that after introducing each new topic or each new reaction, a professor (or tutor) could relate what was just learned back to the core principles needed to understand organic chemistry. This could ground the student in a stronger foundation of what is happening and also they will be able to build onto that foundation as the course goes along. If the structure in the beginning of the course is structured well and outlined as far as the core principles that students need to understand in order to tackle any problem in the class, then the student will have a deeper appreciation for what is going on at the molecular level when a new unfamiliar reaction is introduced.

Conversely, if this type of approach is not taken, like it currently is not, then when a new reaction is introduced it looks completely unrelated to what was previously taught. The student does not directly see the connection between topics. Teaching a new reaction becomes like teaching a new language instead of it being a sense of building on to what was just learned.

So, some of the main concepts that could be related to the majority of organic chemistry problems or steps in a given reaction are:

1. Identify the acid or base in the reaction or mechanistic step (if there is one).
2. Identify the nucleophile or electrophile in the reaction or mechanistic step (if there is one).
3. Identify the leaving group in a particular step of the mechanism or reaction (if there is one). Also, identify at which stage the leaving group will leave. This will determine whether the mechanism follows the path of an S_N1 or S_N2 reaction for example.
4. Are there any stability issues to take into consideration when determining how this molecule will react next? (Such as the presence of carbocations, carbanions, cations, anions, rearrangements, etc.)

If a student learns to identify these type of things with each new topic or reaction that they learn from the start of the course, then eventually they will ask themselves these questions on their own without being prompted by a professor or tutor. The student will have been trained to think about a new roblem in the right way instead of being trained to memorize overall reactions. They will become their own teacher at a certain point when it comes to trying to understand reactivity of new foreign reaction. A student that learns from an early stage how to ask the right questions of themselves, and the reactions in front of them, will be more successful because they re an active participant in their learning and they know what to look for while being taught. If a student can question themselves, they can deduce what a reaction product will be without having to have memorized the reaction in advance. This strategy minimizes the amount of rote memorization needed to pass the course and puts more emphasis on truly understanding basics and allowing those basics to guide your understanding of newly encountered problems.

The student essentially goes from metaphorically doing problems with their eyes closed, to doing problems with their eyes open. The tasks become easier when you develop your foundation first.

PART IV: EXAMPLES OF ORGANIC MOLECULES IN PLACES THAT YOU ENCOUNTER DAILY

EXAMPLE 1: NATURAL BODY STIMULANTS

Do you really think you need that third cup of coffee? Th caffeine provides an energy boost, to be sure, but that artifici stimulant is nothing compared to some of the natural organ compounds produced in your body every day.

That coffee might help you get through another hour study. But what helps you feel the way you feel and want th things you want? Your own body's chemistry is constantly work producing satisfaction, yearnings, and desires. Sometime barely perceptible and sometimes profoundly so, your life is never-ending emotional roller coaster of chemically induce stimulation.

When you are together with someone you care deep about, with whom you share a deep and abiding emotional bon your bodies are communicating with each other as actively a are your words and ideas. The bodies of intimate lovers sta communicating as soon as they *think* of each other, let alor come in physical contact with each other.

That weakness in your knees, that light-headedness, that fluttering sensation and quickening pulse rate are all the effects of an internal release of dopamine in your brain. It is the cause for those suddenly soft Puppy-dog eyes.

Together with norepinephrine, dopamine is one of the important "first response" chemicals called monoamines, essentially brain neurotransmitters that activate all those feelings you associate with falling in love. It is this feeling of euphoria that is so rigorously sought in artificial chemically induced states from a wide range of pharmaceuticals and other sources, which are routinely abused for non-medical reasons.

Norepinephrine   Dopamine

Of course, your body doesn't stop there. The initial responses of butterflies, sweaty palms, increased heart rate, and that subtle confusion that seems to make those simple verbal communications so challenging, are all attributable to either or both dopamine and norepinephrine. But this euphoria would soon pass if it weren't for reinforcements arriving on the scene.

When those first few moments of euphoria begin to flag, a collection of reinforcing chemicals kick in to continue the roller-coaster ride. They include testosterone and pheromones. The latter are still under study in humans, but it is acknowledged that pheromones are heavily involved in interactions in the insect world.

Testosterone is a major player, and it's not just for males, either. Dopamine may start the engine, but testosterone flies the plane. But that's just what happens when you are stimulated in love situations. Your study session is another matter.

Many of the same principles hold true for studying. Natural body chemicals released in your blood enhance your capability to maintain concentration over time. The trick is in knowing how to trigger their release.

Endorphins are a group of "feel-good" polypeptides that combine with some of those same neurotransmitters in your brain, to produce some of the same effects found in the love-reaction processes described above. Some of them create a very similar sense of euphoria. Others relieve pain. Still others are active stress reducers. Their release into your bloodstream is induced by physical exercise.

Exercise increases your metabolic rate, warding off fatigue. Metabolite levels in your blood are perceptibly increased after only ten minutes of moderate exercise. And the added oxygenation of your blood through increased respiration of fresh air can trigger all sorts of good things. Oxygen makes almost everything in your body work better.

So, if you think you need that extra cup of java, why not try just standing up, stretching, and taking a brief walk in the fresh air. You might be surprised at what your own body can do to revive your study session's productivity.

Your body's internal chemical release mechanisms may be the only stimulants you need.

EXAMPLE 2: THE PHEROMONE CONNECTION

Pheromones are organic chemicals within the body that are released to attract the attention of another person. Mothers release them to draw infants closer, just as a woman releases them to draw her partner to her during times of arousal. Unlike regular hormones, pheromones act outside the human body, often as mild scents that gain the attention of the desired person.

Sexual partners, who have been in a physical relationship for some time, know their mate is close by. Their sense of attraction becomes more intense if their partner is sexually aroused. When a man or woman sees the object of their attraction (hopefully, their mate), their body begins to produce pheromones. Once the pheromones are released subtle changes in body chemistry begin to produce the aroma or musk that is used to draw the other person close.

Androstenone
(one type of pheromone)

Testosterone

Notice the structural similarities of the pheromone to testosterone ndividuals who are not in a committed relationship or have just met may also be affected by the subtleties of pheromones. When physical attraction occurs, the level of pheromones produced by the body increases dramatically. The art of attraction is a part of the mating process and, in most cases, pheromones play a fairly large role when it comes to the extent of a person's sexual arousal. Just as reproductive hormones lead to the body's physical reaction, pheromones play on the person's mental and emotional state.

Individuals are capable of picking up another person's scent, just like animals in the wild. Once the scent is detected, chemicals in the brain begin to trigger responses in the brain that lead to, not only curiosity, but exploration and desire. The more pheromones that are released, the more intense the desire to find and connect with the person responsible.

Are pheromones responsible for love at first sight? Probably not, but they have everything to do with the physical attraction two people feel towards one another. Whether they have just met or have been married for 50 years, the butterflies in the pit of the stomach feeling they get when they are together is a result of the affect pheromones have on the body.

Pheromones are the ectohormones that attract one human being to another, be it a mother and a child or two lovers.

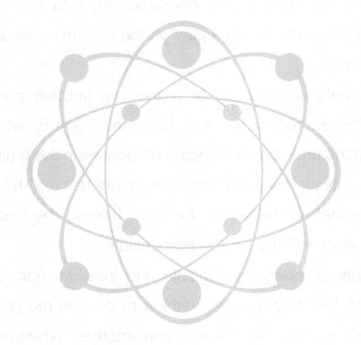

EXAMPLE 3: SUPPLEMENTS

ALA
18 carbons

EPA
20 carbons

DHA
22 carbons

What do you think of when you read on some food label, the acronyms DHA, EPA, and ALA? When I (or maybe you) think about omega-3 fatty acids, I (or maybe you) think about heart health and how much should I have on a regular basis, what foods contain them, and any other questions that have nothing to do with organic chemistry; completely understandable. If I had not taken organic chemistry, I probably wouldn't have cared less about what the structure of these look like. However, if we look at the structure of an omega-3 fatty acid we can begin to see the organic chemistry.

The structure has the functional groups such as alkenes (double bonds between two carbon atoms) and carboxylic acids. Now, it's important to notice that there are multiple double bonds in the structure.

Vitamin C

Vitamin C is another molecule that you hear mentioned often, but probably don't give much thought to; I know that I usually didn't. You probably know some things like how orange juice is a good source of vitamin C or even that vitamin C is essential for the immune system. Kiwi, along with some other fruits and vegetables, is also a good natural source of vitamin C and I have heard it recommended that if you can get your vitamin C from a natural source, then buying supplements could be a waste of your hard earned money.

Did you know that Dr. Linus Pauling, a chemist and two-time Nobel Prize winner, advocated the eating of GRAMS of vitamin C…GRAMS! The FDA recommendation for a daily intake is only on the milligram level, which is a much smaller amount. The amount of vitamin C Linus Pauling was taking could possibly have been killing him slowly because I have seen some research that suggests ascorbic acid could potentially lead to DNA damage. Linus Pauling eventually died of cancer in 1994. So while it seems that small amounts of vitamin C in your diet has research supporting beneficial effects, too much of a good thing can still be a bad thing. But if you are just drinking lots of orange juice and eating fruits, then you have nothing to worry about.

Not many people know the molecular structure of vitamin C though so if for some reason you end up remembering the picture shown here, then you might even get some cool points for drawing it on a napkin at IHOP® while you are eating breakfast with your friends. Or they might just think you're a nerd and send a couple of jokes your way before changing the topic because they have nothing intelligent to add to the conversation about molecular structure.

All of these things aside, do recognize the different functionalities on vitamin C such as the hydroxyl groups, carbon double bonded to carbon, and carbon double bonded to oxygen (amongst some other features that I will leave unmentioned).

The ability to notice particular arrangements of bonds will become important during your time in organic chemistry.

EXAMPLE 4: THE CAFFEINE KICK AND OTHER CONSEQUENCES

I know a large number of people that worship places like Starbucks® or any other good coffee shop. Now whether you like coffee or you like tea, they both (some teas, but not all) contain caffeine in them.

Caffeine   vs   Adrenaline

Caffeine and adrenaline look structurally different, but have similar activities in the body because they share cyclic AMP in their mechanisms of activity. Caffeine blocks the breakdown of cyclic AMP which means that the concentration of cyclic AMP in the body stays up and you will get the same effects that adrenaline gives you because the concentration of protein kinase A will stay up.

Caffeine is just one of the stimulants consumed in large quantities by students and others seeking an energy boost. It is the most widely used and also perhaps the most powerful of stimulants consumed to enhance alertness.

For the purpose of warding off drowsiness, caffeine is the key element in coffee, tea, cola, and energy drinks, to name a few of the most common sources. Of these, energy drinks present the most problematic picture from a safety standpoint. The problem with these beverages is in part related to their high caffeine content. But they also include a laundry list of other stimulant ingredients in combinations the results of which are not yet fully known.

Cramming for exams sometimes does require an energy boost. But caffeine has more effects than just a boost to your energy level and your ability to study longer. It would probably take more cups of coffee than is practical for most students to reach any sort of danger level. But energy drinks contain far more caffeine than coffee.

Caffeine is known to have several very useful effects for students putting in long hours of intense study. Even for normal daily activity, consuming caffeine in combination with sugars throughout the day is known to help maintain alertness. But intense study can demand more concentration and endurance than normal. It is during this time of increased need that the possibility of reaching dangerous levels of consumption might occur. But dangerous levels could be reached far faster by some

individuals with a greater sensitivity to caffeine, or with a pre
existing physical condition or drug interaction possibility.

Caffeine acts to inhibit a natural body chemica
adenosine, which normally limits the body's release of othe
stimulating neurotransmitters, epinephrine, norepinephrine
dopamine, and serotonin. Not held in check, these stream int
the blood and create an energy high. So, caffeine decrease
fatigue and increases alertness, while increasing the body
metabolic rate.

It can also enhance athletic endurance by mobilizing th
body's fat stores and stimulating muscles to burn that fat. Th
delays the muscle's normal depletion of glycogen, which allow
for longer periods of exercise before muscle fatigue become
critical. But for that reason caffeine is one of the mar
substances banned by the International Olympic Committee, ar
is accepted by other sanctioning bodies only at specific low
blood concentrations.

Caffeine has a few nasty habits, too. It will obviously hav
negative effects on anyone suffering from insomnia. It als
increases blood pressure, has been linked with increase
anxiety, can cause digestive disturbances, reduces fine mot
control (the "shakes"), and has been linked with audito
hallucination.

One of the more troubling side effects is an increase in heart rate and possible irregularities in heart rhythm. Caffeine binds to adenosine receptors in the heart muscle mimicking the effects of epinephrine, which results in a stronger and faster heartbeat.

All of that is bad enough, but it is also addictive, and withdrawal symptoms include a list of ironic effects: Decreased alertness, drowsiness, and headache.

But if too much caffeine can be a bad thing, what about all those other ingredients in energy drinks? Aren't they also stimulants? Yes, and yes. They are only now being studied in these potentially dangerous combinations, and we simply don't yet know what their long-term effects might be. As little as one cup of coffee can produce unwanted side effects in those particularly sensitive to caffeine. So a more potent energy drink could easily cause problems for these individuals.

Energy drinks typically contain blends of taurine, guarana, ginseng, ginkgo biloba, and various forms of sugar. All of these act as stimulants, and in combination with caffeine they can be far stronger and possibly dangerous for some people.

Caution is advised. In a 2010 study, Energy Beverages: Content and Safety, published in Mayo Clinic Proceedings, the authors summarized more than 30 years of published scientific literature on the subject, and concluded that we cannot yet predict the level of danger from the ingredient combinations in these drinks.

Your own body is the best barometer to determine how much it can safely consume of any or all of these ingredients. Mild adverse reactions are warning signs to be heeded. And a bit of exercise might serve you well instead of having another artificial energy boost.

An artificial energy boost is a great thing if you can live with all of the other side effects.

EXAMPLE 5: SUGARS: SINK OR FLOAT?

[Glucose structure]

Glucose

Let's say that you have two soda cans; one Coca-Cola® and one Diet Coke®. If you were to place those two cans in a large container of water, what do you think would happen? I'll tell you if you don't already know; Diet Coke® will float, while Coca-Cola® will sink to the bottom. Feel free to try it if you don't believe me. Now, why do you think that is? If you look at the nutritional facts box on the side of each can you will notice that the 12 ounce Coca-Cola® can contains 39 grams of sugar, while Diet Coke® will not have a significant source of sugar. The sugar content in the Diet Coke® will be so small  that it may not even be listed in the amount per serving box; making the can less dense. If we assume that there is about 1 to 4 grams of sugar in one sugar packet (like the kind you find at a

77

restaurant table), then drinking a can of Coca-Cola® is nearly the equivalent of just taking in around 15 sugar packets (using an average of 2.5 grams per packet to make math easy).

When we eat, our glucose (sugar, structure shown above) concentrations spike up drastically before gradually declining. So let us mix a little organic chemistry with biochemistry here. The body uses glucose, from the food we eat, as a main source of energy (if you've taken biochemistry, think of glycolysis, krebs cycle, and electron transport chain). If you know someone with diabetes, then you may have heard the terms "glucose" or "insulin" before and I'll tell you why. Those two terms are often associated with each other because the binding of insulin to the outside of a cell is what allows the glucose to pass to the inside of cells. People who take insulin injections have to do that because their body is not producing enough or any insulin to help regulate the glucose levels that are in their blood. And if their blood-glucose levels remain high, that is what leads to glycosylated hemoglobin (A_{1C}). Hemoglobin (found in red blood cells) is what our body uses to transport oxygen through the body and if it becomes glycosylated (meaning glucose attaches), then it won't be efficient at transporting oxygen (more specifically, glycosylated hemoglobin sucks at releasing oxygen once bound) parts of the body; which is what leads to amputation of limbs n severe cases.

At some point after we eat, say in a dining hall on campus, ucose transports into our cells and is used up by either exokinase or glucokinase (both in the liver) to begin the process f glycolysis. You don't have to remember these specific names you don't want to, I am merely introducing them to you just in ase you have to take biochemistry at some point or you just ant to know. A kinase is just an enzyme that phosphorylates its ubstrate so if you see a name like glucokinase, then you know phosphorylates glucose. Hexokinase will phosphorylate exose (feel free to Google® what hexose is) and so on.

Keep in mind that a cell will meet its own energy equirements first; so once it has enough energy, the rest (e.g. xcess energy from food we take in) has to go somewhere else. you think of the energy produced from glucose as "energy urrency" (like it was taught to me), then if you already have too uch money in circulation, you don't want to produce any more. o when "energy currency" is high, the body will switch from aking energy to storing energy.

Now, with the extra glucose, the liver stores it as glycogen hich can later be broken down for energy if needed) because at is one way of storing excess energy in addition to storing it as fat. The liver is the only organ responsible for maintaining blood-glucose levels so it has the ability to make as well as degrade glucose. Every minute of every day we are depleting our blood-glucose and using up stored glycogen for glycolysis. And overnight the liver depletes its glucose levels, making breakfast so important for replenishing the glycogen storage.

It's not just soda's that have these high amounts of sugar content; a lot of drinks have them. For instance, in a typical day for me I will drink a glass of Tropicana® orange juice (22 grams of sugars) in the morning, I usually drink a glass of apple juice (28 grams of sugars) with my lunch, and I might have a soda (Sprite® is 38 grams of sugars) with dinner. That is already 88 grams of sugar that I've taken in for a typical day; not including if I get thirsty outside of those three meal times. Also not including any sugars that have gone into preparation of the food I eat.

I am mentioning these things to you so that you are aware of what you are putting into your body. Additionally, I hope that when you think of sugar, you can visualize the functional groups (such as hydroxyls) on glucose as well as consider some of the chemistry in your body that happens once glucose is introduced through the everyday tasks of eating and drinking.

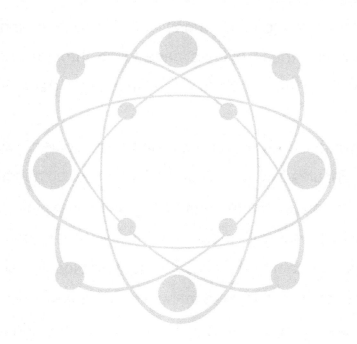

EXAMPLE 6: LETS HIT THE GYM

Being that I am in school at Clemson University right now, the focus on football in this area is ridiculous. A lot of school spirit is shown by the home crowd at football games or even the "solid orange" Friday's when most of the students and faculty wear orange. Watching the football games leads me to think about the amount of training these players have to go through and also how crowded the campus gym is when I'm trying to get a workout in.

The human body is full of organic molecules, such as biological molecules like amino acids which make up things such as peptide hormones and different proteins. If you go into a store such as GNC® you will more than likely see performance supplements that contain biological organic molecules such as leucine, isoleucine, valine, glutamine and arginine. I am using the picture shown here not because I am endorsing it, but because this just happens to be what I use at home. So, the question becomes why are those molecules important to athletes and individuals that workout regularly?

You say you enjoy exercise? Maybe in preparation for better endurance in athletic competition? Skeletal muscles are essential when the body moves so they help the bones as well as joints. Let's take a look at what's in those skeletal muscles, and how you can build them stronger and faster.

About 40% of the average person's body weight is muscle, and muscle is largely composed of protein. Building stronger muscles involves building protein. But the body builds its proteins largely from amino acids, which are in turn the key to building stronger muscles as well as better endurance for athletic competition.

Muscles tend to fatigue. That's what limits our ability to do maximum work levels for sustained periods of time, and it's also what limits our ability to sustain lower levels of work output over longer periods of time. The two activities and the two timeframes are equivalent to the difference between a sprint race and an endurance race. In every kind of muscular activity

83

from body building to endurance competition, amino acids promote the processes of fat loss, strength gains, energy release, muscle recovery, and muscle tissue enlargement.

$$
\begin{array}{ccc}
& H & \\
O \diagup C \diagdown O & \\
H \diagdown N-C-H & \\
H \diagup & C-H \\
& H_3C \quad CH_3
\end{array}
$$

Valine   Leucine   Isoleucine

At peak performance levels, muscles fatigue more quickly. But in endurance competition, such as running a marathon, your body can be conditioned to operate more efficiently, over longer and longer periods of time, effectively postponing fatigue. To a lesser extent endurance training will also increase the muscle's ability to perform at peak levels for slightly longer periods, as well.

One of the most interesting facets of amino acid uses in our bodies is that our bodies effectively "decide" how to use them. Amino acids are withdrawn from the blood by many different bodily processes on a need basis. Our body's first priority is always to maintain sufficient energy to carry on vital functions such as circulation, digestion, and respiration. But

ith adequate levels of dietary fats and carbohydrates, the
ody will refrain from breaking down proteins for energy use,
ermitting their use for other functions and allowing protein
ynthesis to proceed. In other words, muscle can then proceed
o be built.

Glutamine

Arginine

Amino acids are also classed in two groups, essential
nd non-essential. If a non-essential amino acid is needed, our
odies can actually manufacture it from other amino acids. If
ere is an inadequate supply of any essential amino acid, any
ctive processes requiring it cease. Some of the most active
mino acids in muscle development are the non-essential
omponents, arginine and glutamine, and the essential amino
cids leucine, isoleucine, and valine. All of these work together
 muscle protein production.

One of the most interesting body processes for the development of endurance and overall muscle development is the use of the three essential components, leucine, isoleucine, and valine, known as branched chain amino acids. These are directly metabolized within the muscle, converting to energy and thus slowing or even preventing the normal process of ongoing muscle breakdown.

Now, glutamine and arginine are essential for immune system function. Both of which get consumed by two of the major immune cells (lymphocytes and macrophages) at a high rate and because of this, overuse of the muscles (in exercise) contributes to decreased lymphocyte function due to the overall protein and amino acid loss. Arginine was shown to increase the activity of natural killer cell activation, lymphocyte reactivity and lymphocyte activation of natural killer cells in patients with breast cancer. Arginine also lowers the amount of cell adhesion molecules in native cells (thereby thwarting viral and bacterial entry) and by lowering pro-inflammatory cytokines.

Exercise, aging, and illness all have one thing in common; they increase the body's demand for essential amino acids and non-essential amino acids. While the muscles contain the richest storage of amino acids in the body, amino acids are also used predominantly in brain metabolism, neurotransmission, gastrointestinal health, immune function and cardiovascular health. Amino acids can also contribute to energy production through their conversion into glucose via gluconeogenesis.

A study by Ohtani, Sugita, and Maruyama concluded that supplementation of these amino acids prior to, during, and immediately after exercise can provide a source of energy, prevent muscle breakdown, and even speed muscle strength recovery after exercise. What this means is that you can orally supplement these essential amino acids before, during, and immediately following exercise to improve the efficiency of your exercise program.

Your exercise is your method for improving muscular strength and endurance. Amino acids are the tools you can use to fine tune and optimize that program. But all you have to do is provide your body with the tools, and it will handle the details for you.

Amino acids are the building blocks of the protein in your muscles. You can strengthen those muscles faster, increase endurance, and speed muscle recovery with amino acid supplementation.

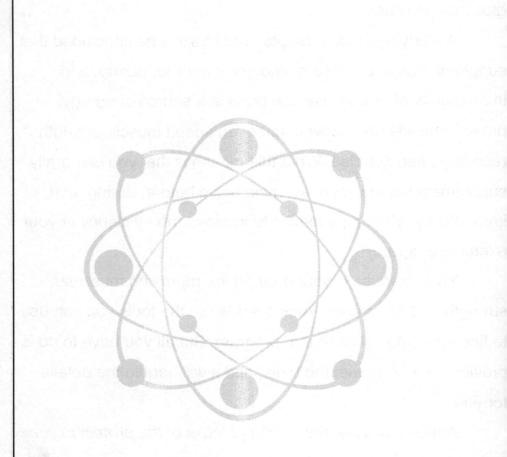

EXAMPLE 7: WEEKEND DRINKING

$$H-\overset{\displaystyle H}{\underset{\displaystyle H}{C}}-\overset{\displaystyle H}{\underset{\displaystyle H}{C}}-O-H$$

Ethanol

I know that for college students (21 and older of course), the weekend can be a time for hanging out with friends and grabbing some beers or even some liquor.

Having a few drinks over the weekend with your friends is almost a rite of passage for a college student. This past time is just a harmless activity that lets busy and stressed college students blow off a little steam after a challenging and grueling week of group projects, late night cramming sessions and little sleep, right? As much as you might want to believe that drinking every weekend is an innocuous way to have fun with your friends, the reality is quite different.

The way your body interacts with the alcohol you drink is determined, in part, by the amount of alcohol present in the beverage you choose. Not surprising, liquor tends to have the highest amounts of ethyl alcohol, or ethanol, the key ingredient in drinks that helps you relax and lower your inhibitions. Beer is at the lower end of the spectrum, and traditional wine is somewhere in between the two. Ethanol is also one of the ingredients that is used to make the fuel you put in your car on the way to class.

Regardless of what type of alcohol you consume, in less than a minute, the effects on your body can be noticed. It passes rapidly through the walls of your stomach and begins to affect your brain. As it courses through your body, the alcohol is broken down into several components that are easier for it to digest. One such component is acetaldehyde, which is made in your body by an enzyme called alcohol dehydrogenase.

The presence of this enzyme in your body has a host of side effects, some of them pretty unpleasant. Headaches, nausea and vomiting are some of the more common of these side effects. Acetaldehyde is actually toxic to your body, just as ethanol is, though acetaldehyde is much more so.

Acetaldehyde

When your body is attempting to metabolize the alcohol and acetaldehyde, it uses NAD+, a compound that is also used keep your cells healthy. Once your body uses its allotment of NAD+ for that day, it has to make more. Your cells convert a compound called pyruvate into energy which the body then converts to lactase. Long term lactase build up can result in a compromise in your liver function as well as kidney damage.

Acetaldehyde can also cause long term damage to your DNA. It does this by binding itself to DNA and changing it which can raise your risks for developing cancer. Acetaldehyde also binds itself to a protein called glutathione. When this happens, the glutathione cannot perform its job inside the cell which can lead to damage.

You can prevent many of the side effects of drinking, such as a hangover, by drinking only while you are eating, drinking plenty of water and drinking only in moderation.

By getting drunk every weekend, you might actually be poisoning your body as it tries to rid itself of the ethanol that is washing through it.

EXAMPLE 8: HAPPY PEOPLE LIVE LONGER

The world's oldest woman (in 1997), 122-year-old Jeanne Louise Calment, attributed her long lifespan to a diet of wine and chocolate rather than one of restrictions. Sarah Knauss, the next oldest person at 119 (died 1999), was said to owe her longevity to being an extremely calm and tranquil individual. Similarly, Christian Mortensen, the world's oldest man, claimed that his good health was due to staying positive. Throughout the history of medical science, there have been many scientific claims of dietary adjustments, sleeping regimens and other habits that could contribute to living longer. Among all of them, happiness is one of the most commonly recurring.

Do Happy People Live Longer?

A study in 2007 found that happiness overwhelmingly reduced the risk of coronary disease in 6,000 participants. At the other end of the spectrum, many studies have shown that depression and negative emotions will eventually increase a person's chances for diabetes, stroke and other diseases.

One study of 3,800 United Kingdom citizens found that those who were happy had a 35 percent lower risk of dying than the other patients. Similar studies have found that happier individuals that are diagnosed with cancer tend to have better results than depressed individuals with similar diseases. All of these studies seem to point to there being a clear correlation between mood and health.

Why Do Happy People Live Longer?

There are two major possibilities that these studies could imply. The first possibility is that happiness releases certain hormones or chemicals into the blood stream, such as cortisol. These chemicals then work as natural antidepressants, bolstering the individual's mood further and causing significant changes in the individual's body. These chemicals could lower the person's blood pressure and decrease the person's tension, all of which can lead to better health overall.

However, scientists also need to recognize that there is a second possibility. The second possibility could simply be that people tend to be happier when they are healthier. Individuals may subconsciously notice the signs of illness long before they have actually been diagnosed as sick and could be less happy as a result of knowing they're sick. Similarly, individuals that have good health and feel as though they will remain in good health for a long time may be happier than others naturally.

Do happy people live longer? Many studies indicate that happiness might be a hidden path to longevity.

Cortisol

Does this structure above look similar (not the same) to those seen in example 2? It should...

EXAMPLE 9: HUGS ARE GOOD FOR YOU

Everyone experiences stress, and too much of it can have a negative effect on your health. Both your physical health and your emotional health can suffer. No matter what is causing you stress, though, from relationship problems to schoolwork to trying to balancing it all, it is nothing that a little hugging can't fix.

Sound too good to be true? You might be surprised to learn that there are important chemical reactions that go on in the body when we are hugged. Hugs release oxytocin, which is known for reducing cortisol. Cortisol is the hormone that our bodies release when things stress us out which is why it's often called the stress hormone.

Dopamine, also known as the pleasure, or happy, hormone, is also released when we are hugged. This helps us to see the brighter side of life, and can make us realize that things will get better.

The success of the hugging movements that are sweeping across the nation attest to the fact that this type of physical contact is not only something our bodies need, but something that we, as social beings, crave. From people who bill themselves as "Professional Huggers" to those that offer free hugs in public spaces, there is something about the touch of another human being that sets off a positive chemical reaction in our bodies that is vital to our health.

No one is immune to the requirements for hugs. Even men, long known for being too tough to need physical contact beyond an occasional punch on the arm, have recently been shown to fare better when they receive regular human touch of a platonic nature. It can't be just any kind of touch, though. It needs to be gentle touch, and hugs are the perfect vehicle for such touch.

They don't just feel good! Hugs are good for your health, too!

Organic chemistry can be found very easily in topics relating to health because a range of chemical reactions occur within the body all the time. A lot of those reactions occur because of the functional groups present on molecules and how they react with other functional groups on additional molecules.

A Challenge to You:

Think about what topics interest you the most for a minute or two. It doesn't have to be science related at all or it can be; completely up to you. Maybe even write it down on paper, leaving space between topics. For example, the first things that came to my mind just now are movies, music, food, relationships, and playing sports/being active. Notice how even though chemistry is what I have degrees in, it was not one of the first things that come to my mind. Now it's important to note that you don't have to make an interesting connection between chemistry and all of the things you write down, maybe only a couple or a few. Even though I love music and it made it on my list, I might have a harder time trying to make that connection to chemistry. As far as movies, there is plenty of chemistry that goes into special effects like explosions, but I know that is not where my passion is. The point here is that if you dig enough, you will be able to find chemistry in any topic but that doesn't mean that what you find will make you passionate about learning organic chemistry; which is one point of this eBook.

I'll choose one of the things from my list, food, because I am a natural fat kid at heart (no offense intended to anyone). I know there are numerous books and YouTube videos and courses that talk about food chemistry. If you're interested in that topic, I suggest looking further into molecular gastronomy. To me, it is very interesting stuff and it might be to you also if you are curious about things like the chemistry of chocolate, how cheese is made, why the form of eggs change upon heating, or anything that involves flavors or cooking. If at some point you think you may want to be a chef and you are also interested in science, then you should definitely look into molecular gastronomy. I don't want to focus on that topic too much here in my example for a couple of reasons 1) there are lots of references that can be found on that topic and 2) while it's interesting to me, I know it's not my passion and it would not make me want to pay attention in an organic chemistry lecture. That might be different for you though so knowing yourself is definitely important. You could definitely throw facts learned about food chemistry into a random conversation with friends in a dining hall or even with family around thanksgiving. Your family might even appreciate it because they will see their college tuition dollars at work through the knowledge you share.

Moving on to another topic on my list, relationships. I can talk about relationships for hours and hours if I have someone willing to listen; whether it's relationships that I have been a part of or relationships that other people are in. I'm sure that I'm not alone in this because I know a lot of people that like to gossip about other people's relationships or won't shut up about their own relationships. Anyway, even in relationships you can find chemistry. Now, when I am thinking about chemistry and relationships I'm not thinking about that mystical feeling between two people that is either there or it isn't; I'm referring to chemical changes that happen inside the body of each individual. Let's say that you just met someone new and you two really start to like one another. That feeling of happiness that you may feel upon seeing a text from them, hearing their voice, or even when they walk in the room; there are chemical changes happening in your body. If your relationship gets to a point maybe past infatuation stages and say your partner starts to stress you out at times or you break up and you become sad; chemical changes are occurring inside of you that you may not even know about. Many organic molecules are present inside of us all and they travel through us and interact with other molecules with causes us to do things like sweat or make our mouth water or make our stomach "growl" or even urinate.

The final topic on my original list was playing sports/being active. Ever since I can remember, I have always enjoyed going outside and playing with my friends. However, it wasn't until I started dating this girl who was into nutrition and fitness that I began to think about the chemistry that happens during exercise, the chemistry that goes into sports drinks and formulas, and all things related. Side note, as a man, I realized over the years that women can make you get interested in things that you never thought you would be interested in. Back to the point though, when I looked at all the deep connections that I made to chemistry in my own time, I noticed that they all involved the human body. With that realization, naturally I was drawn to learn about various biochemistry topics that interested me; even though not all biochemistry interests me. But the point is that I found out what it was that I liked the most and once I took classes like organic chemistry, I tried to rationalize how it might be used in the body or how people use it to transform naturally occurring organic molecules.

Being that I have found ways to make connections between chemistry and relationships or chemistry and exercise, I have been better able to maintain an interest in learning organic chemistry. Because overall, organic chemistry deals with making transformations to organic molecules and it's in my mind that if I can use the tools given to me in these classes then I can take some biologically relevant organic molecule and make a useful modification in an attempt to maybe save lives one day. So, in my mind, the more transformations I am shown and learn, the more tools I will have at my disposal when trying to design some useful pharmaceutical.

There is someone in everyone's family that has some type of health problem no matter how big or small the problem might be. We are all human and we all get sick at some point in time whether it is a common cold or diabetes or some form of cancer. So, to me, everyone could benefit from knowing a little organic chemistry.

Not only that, but typically in organic chemistry courses, polymers aren't discussed very much even though it is probably like a trillion-dollar industry. I am 100% sure that everyone encounters polymers in some form in their everyday lives, yet we don't teach students about it. Things like water bottles, Tupperware, tires, clothing, are all made from polymers. In the body, DNA, RNA, peptides, proteins, are all biological polymers.

My challenge to you (should you choose to accept it) is to do a little digging around on the internet and relate the topics on your own list to chemistry in a way that makes you excited about the random facts you've found. And I mean excited to the point where you can't wait until you have a conversation with someone and you can easily slide in the new random facts you've learned. Once you've made that connection, you will want to learn more that you can use. It may even serve as a conversation starter in some situations when you don't know what to say to someone you're interested in, you never know. The point of this is that you have the power to make chemistry interesting to yourself regardless of whether or not your professor makes organic chemistry interesting in the classroom.

One other thing, while going through the course, try to define the basic concepts early into the course so that you can relate all new information back to them throughout the course. If you need help doing this, your professor should be able to guide you. Ask questions like are any of these molecules an acid? What is the electrophile in this reaction? What position on this molecule is the most electrophilic? What part of my nucleophile is the most nucleophilic? Once the nucleophile attacks the electrophile, are any bonds broken or was the nucleophile attacking a carbocation? Once you learn the right questions to ask yourself, you will reduce any level of struggles you are having with understanding organic chemistry.

About the Author

I have a Bachelor of Science in chemistry and I also received my Masters of Science and Doctorate in chemistry with a concentration in organic chemistry.

In my years spent in universities, I have tutored numerous students in both semesters of organic chemistry, I have taught a few organic chemistry lectures for my advisor, I have taught over 36 organic chemistry labs (>700 students total), and I have also tutored students in biochemistry. I mention my background because I think it's important to show that I have had a good amount of contact with students taking organic chemistry and being that I am closer in age to them, they have openly shared how they feel while I may be explaining topics to them. It's been very interesting to me to see how different students will interact with a subject when introduced to the exact same topics at the same time.

I want to take this opportunity to thank you for reading this guide! I hope this guide was helpful to you in looking at (and thinking about) organic chemistry somewhat differently. Be sure that you and your friends follow me on twitter @ParisLHamilton. Please, rate and review this guide on

Amazon so that I can utilize your comments and suggestions for future editions.

Also, like the Facebook page Success in Organic Chemistry.

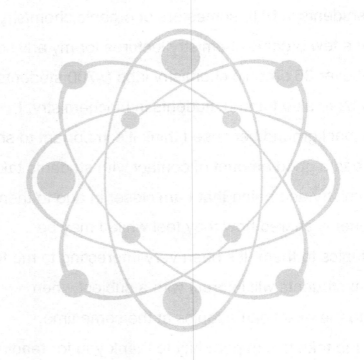

Study Guides and Practice Problems:

More Quotes That I Like

"You can teach a student a lesson for a day; but if you can teach him to learn by creating curiosity, he will continue the learning process as long as he lives."
~Clay P. Bedford

"A teacher who is attempting to teach without inspiring the pupil with a desire to learn is hammering on a cold iron."
~Horace Mann

"I have no special talent. I am only passionately curious." ~Albert Einstein

"Success consists of going from failure to failure without loss of enthusiasm."
~Winston Churchill

"Study while others are sleeping; work while others are loafing; prepare while others are playing; and dream while others are wishing." ~William Arthur Ward

"Believe in yourself and all that you are. Know that there is something inside you that is greater than any obstacle."
~Christian D. Larson

Made in the USA
Las Vegas, NV
28 January 2024

85035319R10059